REALITY SIMPLIFIED

A New Worldview of How Reality Is Here, Changes, and Continues

Jack Wiley, Ph.D.

Jack Wiley Publications

http://www.amazon.com/author/jackwileypublications

ISBN: 9781508732983

Updated 11-16-23 Version

CONTENTS

CONTENTS

1

INTRODUCTION

We are here and it is now. Further than that, all human knowledge is moonshine.
~H. L. Mencken

"We are here and it is now" has always been true for me. But I question Mencken's "Further than that, all human knowledge is moonshine." My background in science leads me to believe that most of the knowledge gained by the so-called scientific method is more than just moonshine. However, I would agree with Mencken's quote if the last sentence were revised to: "The further we go from the here and now, the more so-called human knowledge strays toward nonsense." And in my mind this includes so-called scientific knowledge.

The present most accepted worldview that is embedded both in our daily lives and in science is matter in motion in space and time. English language and most languages of the world support this worldview to the hilt. Immanuel Kant (1724-1804) contended that it is impossible for us to do any thinking about real objects without the concepts of time and space. I disagree. A better and certainly simpler way is to view reality in the here and now as what it is formed into, which is the focus of what this book is about.

Starting in a high school physics class in the mid-1950s, if not before, I was indoctrinated into thinking that Newton's laws of motion described the way reality was really here and worked. Only gradually did I began to question the motion in space "thing" as the way, or even a way, that reality was actually working. It wasn't until 1993, however, when I took a closer look at automobiles (and everything else including myself when I was doing what we say is walking or running or jumping) that were supposedly in motion in space and time that I glimpsed that what I had been conditioned or brainwashed into thinking I was observing was not what I was actually observing. Suddenly I was not only questioning the idea that things really do move in space and time and replacing it with the idea or observation that we observe what reality is formed into and changes that take place thereof instead.

The space and time view has to be right (right?). Otherwise, Galileo and Newton and even Einstein were "fooled." Here's how I resolved this years later in my own mind. When people thought they were observing (and measuring) objects moving in space and time, what they were actually observing and measuring (my opinion) were changes in what the one continuing reality is formed into from what they remembered it (or measured it) to have been formed into on-going over and over again.

In other words, I rejected the motion in space as a way that reality was working completely. That was when I finally realized that I was observing changes or switches in what reality was formed into instead. What I observed seemed both magical and surreal. I suddenly realized that I was observing reality itself "working." From years of daily living and my background in science, I thought I knew something about how things worked, but even the idea of "reality working" had not occurred to me before.

All I had to go on when I decided to try to figure out how reality works is reality, both my own and that around me, that is here and working. What causes it to work the way it does work was and still is invisible to me. However, what we can do is observe "always" patterns or rules or laws in the way we observe reality to be here and working. This is what I have done over a period of many years to form what I call "always true" reality statements or generalizations. From this, I was able to figure out the general way that reality is here and does change and does continue.

Please understand that I am not rejecting the so-called scientific method, which I still think is the best devised so far for understanding how things work. However, I am rejecting matter in motion in space and time as a myth that was never discovered by the scientific method in the first place. The time that supposedly has a beginning and end was a religious myth before it became a scientific myth.

I devised a new method for doing this. I wrote reality statements, which are sort of like hypothesizes used in science. Then I tested the reality statements out to see if they were true for my own actual reality and that which I observed around me. However, my tests were stringent and unconditional. A reality statement had to be true for all reality that I could possibility observe. If I observed a single exception to a reality statement, I tossed it out and replaced it with a revised or completely new reality statement. I could also "cause" changes to further test my reality statements.

The result of this is that I arrived at four reality statements of how reality always works everywhere that combine together to give us a revolutionary new general overview – which I will refer to as a "worldview" – of how reality *is* here, *does* change, and *does* continue. I think of this new worldview as more than a theory, at least more than a predictive one. The ways that reality always works are certainties. They are unconditional and there are no exceptions.

Because the reforming of what the one continuing reality is already formed into is a key way that reality always works, I call this the "One-Reformable-Reality Worldview."

It is the utility – the usefulness – of the One-Reformable-Reality Worldview that has surprised me the most. It has literally and drastically changed my life for the better, and it can do the same for you. It removes the matter in motion in space and time, which I now consider a myth, from science and billions of years of time and replaces it with the observation that the here and now is all that every exists. Read on to find out how a worldview housed in a human mind can do all this – and more!

This book is based on simple observations that anyone can make. Regardless of whether you consider yourself a non-scientist or a scientist, or something in between, this book is written so you can understand it. Unlike traditional science, which builds on previous science and every branch has its own frontier, the One-Reformable-Reality Worldview starts with new observations and goes off in a new direction. This eliminates the gobs of references to what has previously been thought and done found in most books of this type. There are no equations in this book. Instead of equations, I make reality statements. It's all stated in words, plain and simple. The main requirement for understanding and using this revolutionary new worldview is a background in daily living, which everyone already has.

This book is intended as an introduction to viewing reality as what it is formed into and as a how-to guide for doing it in this manner and using the One-Reformable-Reality Worldview both for better living of our daily lives and making the world a better place to do that living.

2

A NEW WAY OF VIEWING REALITY

I define "reality" simply as *that which is here*. The dictionaries I have looked up "reality" in give a number of other definitions of reality, but not my definition of reality. I will not be concerned with any of the other definitions of reality. What makes reality *reality* is the fact that it is here.

Throughout this book, it is important to keep in mind that whenever I use the word "reality," I mean *that which is here*, and vice versa.

The only way we ever see reality (defined as *that which is here*) is as what it is formed into. I will refer to what we see as "what reality is formed into."

All the reality I have ever seen has been formed into a mixture of what we commonly call things and space. However, I can only make visual sense out of things, which I suppose is the reason why we refer to them as "sensible things."

What stand out to me about what the reality I see is formed into are specific things, both non-living and living, including ourselves, *and* the arrangement these things are in with each other.

For example, what I see the reality in front of me formed into right now is that bookshelf over there, the specific books that are on it, the table in front of me, that binder, that pen, that person over there, *and* the arrangement these things, including myself observing them, are in with each other.

Don't I see the positions the things have in space rather than the arrangement they are in with each other? My answer is no. I cannot see any totality to space, or for that matter, even if there is a totality to it, so I have no way of observing the position a thing has in space. I observe an area of what reality is formed into right now as being a mixture the mysterious areas that we call space and things. The only way I ever "see" these areas is along with

Sample of reality formed into things and an arrangement that they are in with each other.

specific things, including myself. In other words, I observe the arrangement that things are in with each other directly, and not in terms of space or a combination of space and time.

I realize, of course, that what I am saying is contrary to the space, time, and motion of classical physics that is firmly embedded in our lives and language as being the way that our world really works. I disagree. What I have going for me, however, is that physicists themselves seem to be shifting from the classical space, time, and motion view to the more computer-like quantum mechanics as the way reality is really here and works. I'm now convinced that this is also a better description of how reality is here and working in our daily lives and a more useful way of viewing it than is the traditional matter in motion in space and time.

My new way is to view what reality is formed into directly. Even though our viewing it is always subjective from individual human minds, the physical reality itself is objective, absolute, and what is really here right now, at least as real as it ever gets. In order to understand what I am talking about, it is important that you view actual what reality is formed into – that is actual things and actual arrangements they are in with each other – rather than just imagining some of these.

In my new way of viewing reality, change is defined as any difference in what reality is formed into from what we remember it to have just been formed into. Since individual human memory is involved in remembering what reality was formed into, we obviously will get much more agreement between humans on what reality is formed into right now when two or more humans are together observing and experiencing the same area of what reality is formed into than what they remember it to have been formed into, and even more so the further back, say yesterday or last year, the memories are from.

As I continue to observe what the reality I see in front of me is formed into, I do not notice any change within what the reality I see is formed into from now to now to now from what I remember it to have been formed into. That bookshelf over there, the specific books that are on it, the table in front of me, that binder, that pen, and that person over there appear the same as I remember them to have been, *and* the arrangement these, including myself, are in with each other that reality is formed into does not appear to change.

My memory plays an important role in my ability to observe this. It requires memory of what I have seen to compare with what I see to be able to observe that what reality is formed into that I am looking at is staying the same.

As I continue to observe what the reality in front of me is formed into, I can recognize change (if there are any) in what reality is formed into from what I remember it to have just been formed into. For example, that person over there that I remember was sitting is now standing. My memory plays an important role in my ability to observe this. Without memory of what I have seen to compare with what I see, I would not be aware of changes that have taken place or are taking place in what reality is formed into.

Noticeable changes in what reality is formed into from what I remember it to have just been formed into are the exception rather than the rule. Most what reality is formed into that I observe stays the same as I remember it to have just been. Even when there is a noticeable change in what reality is formed into from what I remember it to have just been formed into; most of what reality is formed into that I observe along with the change usually appears to stay the same as I remember it to have just been. For example, when I observed that the person I remembered to have been sitting was standing, this was the only change I observed in what the reality was formed into that I was looking at. All the other things appeared just as I remembered them to have been, and the arrangement that these things were in with each other did not appear to have changed.

All of the observable differences in what reality is formed into from what we remember it to have been formed into are in things themselves; or in the arrangement that the things, including ourselves, are in with each other, or in some combination of the things themselves and the arrangements. In other words, all of the observable changes are in what reality is formed into from what we remember it to have been formed into, and we observe these changes in terms of sensible things.

What is strange is that all of these observable differences take place as "switches" in what we remember the reality to have been formed into to what it is formed into right now. What it has switched from is no longer reality.

In order to save words by not having to repeat "what reality is formed into" over and over again, I will call this the reality formation and use the RF abbreviation for this throughout the remainder of this book.

If you have had any science classes at school, and especially a physics course, you were probably taught to view things happening and our lives taking place in terms of space and time. I have read in more than one supposedly science book that this is the "only way that humans have of making sense of the world." My new way of viewing reality – that is, to view it simply as what it is formed into or as the RF – shows that there is another way that doesn't even involve space and time. That way is to simply observe the RF itself is formed into and follow the reforming of the RF as it takes place. This is easy because there is always just one RF.

3

THE WAY ALL
CHANGES TAKE PLACE

An amazing thing to notice about the way changes take place within what reality is formed into is that *as any change takes place anywhere within what reality is formed into, the changed part of what reality is formed into blends together with that which did not change and becomes what reality is formed into*. This has been true for every particular change I have observed anywhere in what reality is formed into, regardless of whether the change was of natural or human cause. Because I have never observed an exception to this, I generalized that this would be true for every particular change in what reality is formed into. To save words, I replace "what reality is formed into" and "reality formation" with the RF abbreviation and restate this: **As any change takes place anywhere within the RF, the changed part of the RF blends together with that which did not change and becomes the RF.** I will refer to this as Reality Generalization 1. In later chapters, I will give other reality generalizations. These generalizations hold true unconditionally for all reality that I (and I think we) can possibly observe. In other words, they are much more than mere predictions; they are certainties. We have all heard it said that the only certainties in life are death and taxes. I am adding some more certainties.

Having any change that takes place anywhere in the RF blending together with that which does not change and becoming the RF is an important way that reality always works everywhere. I call a change and the blending back together with that which did not change a "change and continuation cycle" in the RF. The magical part is the blending together with that which did not change. It appears to happen automatically without anything visible to us causing it to happen.

I have already spoken of our natural ability to recognize what reality is formed into (the RF) that we see and to notice changes in the RF as we continue to observe it. If the RF looks different to me in any way from how I remember it to have just looked, I conclude that it has changed.

In order to understand Reality Generalization 1, it is important to observe change within the RF as the change is taking place and to observe it as change within what reality is formed into rather than as what we think is happening. After all, it is the changing of the RF that is really happening, and not what we think is happening. The reality change is objective; what we think is happening is subjective.

As a starting example, I continue to watch a person's facial expression, which I observe as part of a thing that reality is formed into. As I do so, I notice that what the person's facial expression is formed into has changed from what I think of as a frown (what I remember the reality to have been formed into) to what I think of as a smile. I conclude that the RF that I am looking at has changed, and that what it has changed to has become the RF. The facial expression is now formed into what I think of as being a smile, and as I continue to watch the facial expression, any change in the facial expression becomes the facial expression that the person has. And so on. Thus, in the case of the person's facial expression, Reality Generalization 1 holds true. *As any change takes place anywhere within the RF, the changed part of the RF blends together with that which did not change and becomes the RF.*

Notice that Reality Generalization 1 holds true for the smallest noticeable change within the RF. After each smallest noticeable change, the change has already blended in with that which did not change and become the RF. There are a series of smallest noticeable changes in the facial expression as it is changing from a frown to a smile. After each smallest noticeable change, the change has already blended in with that which did not change and become the RF.

Since we cannot observe a change within the RF, such as the changing of a person's facial expression, all at once, our memories play an important role in our awareness that the change did or is taking place. When we watch a change as it is taking place, we commonly say that we are observing it. We do this by seeing and remembering the RF right now. Then we see what the RF is formed into again right now and mentally compare what we remember having seen with what we see now and draw conclusions about what has changed. Then we do this over again. And so on. We have the ability to do this on an on-going basis without being aware that we are doing it. We get an on-going comparison in our minds of what reality was just formed into and what it is now formed into with noticeable differences standing out.

When we shuffle a deck of cards, or shuffle whatever we find around us (what we commonly think of as moving things around), it is easy to observe that Reality Generalization 1 holds true. *As any change takes place anywhere within the RF, the changed part of the RF blends together with that which did not change and becomes the RF.*

Notice that this is true for the smallest noticeable change within the RF. After each smallest noticeable change within the RF, the change in what reality is formed into has already blended together with the parts of what reality is formed into that did not change and has become the RF. Then the change process starts over again.

It is easy to observe that the blending together after each smallest noticeable change takes place when the things are in contact with each other; but more difficult when something is supposedly "moving in space." When I do what we say is dropping a ball, does it really blend back with that which did not change after each smallest noticeable change within the RF when the ball is supposedly moving through space? I'm convinced that the answer is yes. When the ball contacts the floor, it has certainly blended back together with the parts of the RF that did not change from what we remember them to have been formed into.

Whenever we say we are observing something moving or in motion in space, I think what we actually observe are changes taking place within the RF from what we remember it to have just been formed into, with the changed what reality is formed into blending together with that which did not change after each smallest noticeable change within the RF. This was true for the "dropped ball." It is also true for what we commonly say is an automobile in motion. For example, when I stand beside a roadway and watch an automobile that is supposedly in motion, the changes I actually observe are within the RF, which I observe as changes in the arrangement that things are in with each other. The arrangement that the automobile and I are in with each other keeps changing from what I remember it to have just been repeatedly. As the arrangement that the automobile and I are in with each other changes, what it changes to becomes the arrangement we are in with each other. I also notice that the arrangement that the roadway and the automobile are in with each other keeps changing from what I remember it to have just been repeatedly. As this is happening, neither the arrangement that the roadway and myself are in with each other nor the arrangement that the parts of the roadway are in with each other appear to be changing from what I remember them to have been. As the arrangement that the roadway and the automobile are in with each other changes, what it changes to becomes the arrangement they are in with each other. Thus, in the case of the arrangement that the automobile and myself are in with each other and in the case of the arrangement that the roadway and the automobile are in with each other, Reality Statement 1 holds true. *As any change takes place anywhere within the RF, the changed part of the RF blends together with that which did not change and becomes the RF.*

Notice that this is true for the smallest noticeable change within RF, which in the automobile example we observe as change in the arrangement that things are in with each other. After each smallest noticeable change within RF, the change within RF has already blended together with the parts of RF that did not change and has become the RF. Then the change process starts over again.

If all of this sounds too complicated to follow, simply observe all of this as the changes that are taking place within the RF as they are taking place.

Change in the arrangement that two continuing things are in with each other is an easily observed change within the RF. For example, I can easily observe a change in the arrangement that something and me are in with each other from what I remember the arrangement to have been, and I can easily observe a change in the arrangement that two continuing things are in with each other from what I remember the arrangement to have been. It is important to observe this directly as changes within the RF as the changes are actually taking place.

I have the ability to change the arrangement that my body parts are in with each other, which in turn gives me the ability to change the arrangement I and a sidewalk are in with each other and to change the arrangement that things I find around me are in with each other. It is important to view all of this directly as what reality is formed into and as the changing of what it is formed into as the changing is taking place.

As I change the arrangement that my body parts are in with each other, the arrangement that I change them to becomes the arrangement that they are in with each other. Try this yourself so that you can see that it is true. Slowly do what we commonly say is moving or raising your right arm from your side. As the arrangement that your right arm and the rest of your body are in with each other changes, the arrangement that they change to becomes the one they are in with each other. This is a change within the RF. Reality Generalization 1 holds true. *As any change takes place anywhere within the RF, the changed part of the RF blends together with that which did not change and becomes the RF.*

Notice that this is true for the smallest noticeable change in the arrangement that your right arm and the rest of your body are in with each other. There are many noticeable changes in the arrangement that your right arm and the rest of your body are in with each other as the arrangement changes from your arm being at your side to overhead.

The arrangement that two things are in with each other can change in a number of different ways besides the arrangement they are in with each other becoming closer together or further apart. For example, the rotational arrangement that two things are in with each other can change. The internal arrangement that parts of a thing are in with each other (essentially smaller things within a thing) can change. An example of this is the changing of the arrangement a piston in an automobile engine and the motor block are in with each other. As an automobile engine is "running," the arrangement that many parts of the engine are in with each other change in a variety of ways. As these changes are taking place, Reality Generalization 1 holds true. *As any change takes place anywhere within the RF, the changed part of the RF blends together with that which did not change and becomes the RF.*

Many changes we observe within the RF are in things themselves. For example, ingredients that a person puts in a pan and bakes in an oven become a cake. As these changes take place, Reality Generalization 1 holds true. *As any change takes place anywhere within the RF, the changed part of the RF blends together with that which did not change and becomes the RF.*

I can observe that Reality Generalization 1 holds true for everything I can possibly do and for all changes I can possibly observe within the RF from what I remember it to have been formed into.

Reality Generalization 1 has also held true for everything I have observed using microscopes and telescopes. It also holds true for observable waves. For example, as the shape of waves on the surface of water change, the shape that they change to becomes the shape that they have. Reality Generalization 1 holds true. *As any change takes place anywhere within the RF, the changed part of the RF blends together with that which did not change and becomes the RF.*

I have never observed an exception to Reality Generalization 1, and cannot even imagine a possible exception. This is an important way that reality always works everywhere.

4

THE WAY REALITY CONTINUES

A major conclusion I made, which I consider a fact and call Reality Generalization 2, is: ***There is just one continuing and changeable RF.*** By "continuing," I mean in existence in a time sense. Keep in mind that "RF" means reality formation or "what reality is formed into." In order to understand Reality Generalization 2, it is important to keep in mind that I define "reality" as *that which is here*, and that the only way we ever see reality is as what reality is formed into, which is most noticeably by us the specific things along with the arrangement these things are in with each other that it is formed into, which I refer to as the reality formation or RF.

Having just one continuing and changeable what reality is formed into is an important way that reality always works everywhere.

I will now describe some of the particular observations and thinking that led to Reality Generalization 2.

I am aware that I continue to have just one reality. I have just one reality right now and just one reality right now and just one reality right now. There continues to be just one of me. Reality Generalization 2 holds true. *There is just one continuing and changeable RF.*

I can observe that there is just one reality around me right now and just one reality around me right now and just one reality around me right now. Reality Generalization 2 holds true. *There is just one continuing and changeable RF.*

There is no place for more than one RF. One RF completely fills everywhere we can observe. Keep in mind that what we consider space is part of, along with what we consider matter, what fills everywhere we can observe, and not something separate from it.

As I observe change in what reality is formed into as the

changing is taking place, it is easy to observe that there continues to be just one reality. I continue to see just one reality; it is what that one reality is formed into that changes.

A wall clock with a clock face and hands that are "running" continues to have just one reality. The clock hands and clock face are in just one arrangement with each other right now. This continues to be true right now and right now and right now. As the arrangement the clock hands and the clock face are in with each other changes, continuing reality is all the reality that both the clock hands and the clock face have. There continues to be just one clock with just one clock face and just one set of clock hands. This is true regardless of whether or not the clock is running. Reality Generalization 2 holds true. *There is just one continuing and changeable RF.*

I have never observed an exception to Reality Generalization 2, and cannot even imagine a possible exception. I consider it a fact. *There is just one continuing and changeable RF.*

5

THE SAME REALITY
IS USED REPEATEDLY

A major conclusion I made is: **The same reality is used repeatedly to form every RF there ever is.** I call this Reality Generalization 3. In order to understand Reality Generalization 3, it is important to keep in mind that I define "reality" as *that which is here* and that the only way we ever see reality is as what reality is formed into, which I refer to as the RF.

Using the same reality to form every what reality is formed into is an important way that reality always works everywhere.

I will now describe some of the particular observations and thinking that led to Reality Generalization 3.

In order for Reality Generalization 1 (*As any change takes place anywhere within the RF, the changed part of the RF blends together with that which did not change and becomes the RF*) and Reality Generalization 2 (*There is just one continuing and changeable RF*) to both be true, the same reality (defined as that which is here) that is formed into the RF now must switch into and become RF now. In this manner, the same reality, with no reality added or subtracted, is used repeatedly. The same reality that is formed into the RF now has been formed into every RF there has been up to now. I have further concluded that the same reality will be formed into every RF that there will ever be. In other words, the same reality continues to be here and is reusable. It is what the same reality is formed into that can change. In all of my particular observations of changes within the RF, Reality Generalization 3 has held true. *The same reality is used repeatedly to form every RF there ever is.*

We do not look back in time; we look back in RFs.

That the same reality is used repeatedly is obvious. What reality is formed into now changes into what reality is formed into now, which changes into what reality is formed into now, and so on, endlessly. In my mind, this renders all explanations about the creation of reality from nothing such as the Big Bang to nonsense.

It is important not to confuse Reality Generalization 3 with the idea that there is one set of atoms that are used repeatedly to form different things. I observe RF now becoming the RF now, without reality ever returning to a basic set of anything.

In all my particular observations of reality, I have never observed any reality (defined as that which is here) being added to or subtracted from that which is already here. I cannot even imagine how this would be possible.

Another observation that supports Reality Generalization 3 is that only what we find reality already formed into can change or be changed. Whatever it changes into becomes what it is already formed into.

I observe what reality is formed into now changing into and becoming what reality is formed into now in many different ways, including mechanically and chemically. There may be other ways that what reality is formed into changes into and becomes what reality is formed into that we have not yet discovered. Regardless of how the changes take place, Reality Generalization 3 is observed to always hold true. *The same reality is used repeatedly to form every RF there ever is.*

Only the one continuing RF ever exists.

6

CONTINUING REALITY

I now bring in the idea of continuing reality as a replacement for that of passing time.

Continuing reality is an observable, in fact, the only observable there ever is. I am aware of my own continuing reality. I do not know why my reality continues, but I am aware that it is continuing. I can observe continuing reality around me. I do not know why reality around me continues, but I can observe that it is continuing. In other words, I observe and experience continuing reality without knowing what causes reality to continue.

My conclusion that *there is just one continuing and changeable RF* (Reality Generalization 2) led to my realization that what I observe and experience as my own continuing reality and as continuing reality around me and what I think of as the present time or now always correspond. Because this is always true, continuing reality is an observable replacement for the idea of passing time.

Notice that I have concluded in Reality Generalization 2 that there is just one continuing reality. This means there is never any reality besides this.

This also means we can mark off what we think are equal amounts of continuing reality as reality is continuing. To do this, we commonly use a change in what reality is formed into that repeats itself that we can count. For example, sunrises repeat themselves and we can count them. There is a certain amount of continuing reality as a day occurs. If we assume there is the same certain amount of continuing reality as the next day occurs, and that this will be true for every day that occurs after that, then we can use days to mark off what we think are equal amounts of continuing reality as reality is continuing.

The principle is the same when the vibrations of certain atoms are counted. There is a certain amount of continuing reality as the atom vibrates one time. If we assume there is the same certain amount of continuing reality as the next vibration occurs, and that this will be true for every vibration after that, then we can use vibrations to mark off what we think are equal amounts of continuing reality as reality is continuing.

A familiar clock with hands and a clock face changing the arrangement they are in with each other is used to mark off divisions of a day as the day taking place. The clock can also be used to mark off how much continuing reality there is as a certain change in what reality is formed into is taking place. Indirectly, the mechanical clock "counts" the number of swings a pendulum makes back and forth or the number of vibrations of a quartz crystal.

Whenever scientists say they are measuring intervals of time, I am now convinced that what they actually do instead is place memorable marks in continuing reality as it is continuing. Rather than "measured time" starting the scientific revolution, I think it was marking off continuing reality as it was continuing that did it.

It seems to me that scientists have been marking off continuing reality all along. The only change needed is to bring what they say they are doing in line with what they do.

It is important to notice that the marks are always after the reality has already continued. This is the only way it can be marked off.

We can use how many standard units of continuing reality are marked off while a certain change takes place within the RF as a measure of "how fast" that change took place. Likewise, we can use how many standard units of continuing reality there are while we wait for a doctor's appointment as a measure of "how long" we waited. I think this is what we do when we say we are measuring intervals of time.

I no longer think there is passing time. I now think there is continuing reality.

7

THE SAME TIME IS
USED REPEATEDLY

Another major conclusion I made is: *There is just one now time that is used repeatedly to give us the continuing RF we observe and experience.* I call this Reality Generalization 4.

The idea that we are fed new time seems preposterous to me. Instead, I think there is just one now time that is used repeatedly to give us the one continuing and changeable RF that we observe and experience.

How can the same one now time, which is all the time there ever is, be used to give us the one continuing RF we observe and experience? I do not know how it actually works, but here is a greatly simplified model of the way I imagine it might work.

The one now time is everywhere. However, in what we think of as a before now and after now direction, it has a "thickness" of less than what we think of as a trillionth of a second time interval.

Up until recently I imagined that this reusable now time as switching back and forth between two states. I now realize that to say that time (who or what is time?) switches back and forth between two states is rather meaningless when we cannot define what we mean by time in a meaningful way. I now think it is matter and/or space that is switching back and forth between states. There is so-called "natural" switching in reality that makes digital computers along with the internet not only possible but practical.

As I imagine it and evidence supports, the switching back and forth between states is very rapid, more than a trillion times in what I think of as being a second of marked off continuing reality. This is true everywhere. The number of back and forth switches of states is a measure of how much continuing reality there was as a certain change took place in the RF or of how much continuing reality there was while I waited to see a doctor.

The one reusable time can only be in one of two states right now where I am, on the other side of Planet Earth, or anywhere else.

This new view of time replaces the ideas of passing time and intervals of time, whatever those are supposed to be, with that of one switching time that is used repeatedly to give us the one continuing RF we observe and experience.

8

A CHANGE AND
CONTINUATION CYCLE

As any change takes place anywhere within the RF, the changed part of the RF blends together with that which did not change and becomes the RF (Reality Generalization 1) is an important way that reality always works everywhere. I call a change and the blending back together with that which did not change a "change cycle" in what reality is formed into. I now add continuation to this.

I visualize the same reality switching back and forth between two states I described in the previous chapter as also allowing changes to take place within the RF, with one state (the "flip") allowing the change and the other (the "flop") bringing the part of the RF that changed back together with the unchanged into one continuing what reality is formed into. Thus, the switching is back and forth from "flip" to "flop" or "flip↔flop."

Every change within the RF and blending together again with the surrounding reality that did not change must take place in a single back and forth reality switching between the two states. After each back and forth switching, we are back to one what reality is formed into; in other words we are at square one again. I call this a "change and continuation switching cycle." This cycle takes place in what I think of as being less than a trillionth of a second of continuing reality. Every change in what reality is formed into takes place in this manner. A complete change and continuation cycle takes place in less than a trillionth of a second of marked off continuing reality. After each cycle, we are back to one what reality is formed into.

Different amounts of change, from none to the largest possible, can take place in a change and continuation cycle, but the duration of the cycle is always the same. These are local changes in what reality is formed into. This is the only way a change ever takes place in the one continuing RF. In less than a trillionth of a second of marked off continuing reality, the change has taken place and blended in with the surrounding what reality is formed into to complete the cycle and we are back to one what reality is formed into. In other words, this is the way what reality is formed into changes and becomes what reality is formed into.

Notice that all changes are switches within the RF from what we remember (or measured or recorded) the RF to have been. The RF literally switches from what it was formed into (if our memories or measurements or recordings are accurate) to what it is formed into now. The remembered (or measured or recorded) differences have like magic disappeared from the one continuing RF.

Of course, with our normal vision, we only notice a series of change and continuation cycles that add up to a change that is large enough for us to notice. If I take a metal object and hold it as close as I can to a table without letting it touch the table and drop it, a very small noticeable change in what reality is formed into takes place in what reality is formed into very rapidly. This change obviously took place in a fraction of a second of marked off continuing reality. It is equally obvious that the change I observed took place as a series of change and continuation cycles. This is what I observe when I focus in on actual change as it is taking place. My conclusion is that a change and continuation cycle is complete in less than a trillionth of a second of marked off continuing reality, and no change in what reality is formed into has ever taken place in any other way.

It is further noted that less than a trillionth of a second is not only the largest interval of time that ever exists, it is also the only interval of time that ever exists. This is the only interval of time that I am salvaging in my thinking from what I now think of as the myth of matter in motion in space and time or spacetime.

9

THE MAGIC OF BEING HERE, CHANGING, AND CONTINUING

Richard Dawkins book, *The Magic of Reality*, was first published in 2011. I agree about his definition of magic, which I will get to after I get his definition of reality out of the way. According to Dawkins, reality is everything that exists. By my definition of reality, reality is existence itself.

How is the way reality is here, changes, and continues accomplished? We observe and experience and are an integral part of the being here, changing, and continuing, but how this is accomplished is hidden from our viewing it. The slightest change within RF from what we remember it to have just been formed into immediately blends together with that which did not change in such a way that what we think of as things happening and the changes that take place in the one continuing what reality is formed into is one and the same. My previous idea that my life was somehow taking place in space and time is replaced with that it is taking place in and as part of the one continuing RF.

How this is accomplished can seem like some sort of magic trick in the sense that how this is accomplished is hidden from our viewing it. But that it is accomplished is here and now for all human beings to observe and experience.

Dawkins says that magic is a slippery word that is used in three different ways, which he distinguishes as "supernatural magic," "stage magic," and "poetic magic." Like Dawkins, I view reality as "poetic magic."

Regardless of how this "poetic magic" is accomplished, it seems to take place at a quantum level and is more computer-like than mechanical. Reality appears to literally "switch" from what it was formed into to what it becomes formed into. Newton's idea that our world works by having some sort of mechanical continuous motion in space now seems preposterous to me. Everything starts with and depends on what reality is formed into right now.

When I ride a bicycle and stop pedaling, I keep on going. I formerly thought this freewheeling poetic magic took place by some sort of continuing mechanical motion in space; I now view this as taking place as the reforming of what reality is already formed into. My pedaling of the bicycle is reforming some of what reality is already formed into which in turn causes some sort of switching places with the one continuing space that is here along with things in the one continuing what reality is formed into. When I'm actually freewheeling on a bicycle, it feels like I'm switching places with the space in front of me over and over again and that the momentum is in the repeating switching of places with space rather than some sort of continuous motion in space, which I no longer think exists.

Our digital computers can work like they do because of the way reality works, and not the other way around. The "natural computers" contained in each individual human brain works in ways far beyond simple digital switching. The operating and control system for reality itself is unimaginably more complex and sophisticated than that.

The switching back and forth in the one time that is used repeatedly, or however this actually works, on a universal scale requires computerization or whatever beyond human imagination, at least beyond mine. All this appears to take place in time switches of less than a trillionth of a second (and if Michio Kaku, Ph.D. and professor of theoretical physics at the City University of New York, is correct, about 500 trillionth of a second) of continuing reality in what we think of as a before-now after-now direction.

Is there a control center outside of reality that takes care of all of this? Would this be something like the way we control puppets by way of strings controlled by a human being from above or a model railroad from an electrical control center off to the side of the model railroad itself? I think not simply because the "outside of reality" would have to be even more advanced and sophisticated than our world. This is why I think the operating system and control and programming – whatever there might be – is built into or along with the one continuing what reality is formed into.

In any case, a whole lot of activity must be going on in the areas of what reality is formed into that we consider space that we cannot see to give us the changes in what reality is formed into and the continuing RF that we observe and experience. We observe and experience "switches" in what reality is formed into without being able to observe anything "causing" the switching.

An amazing thing about this switching is how precise it is. It works the same way always and everywhere without ever missing a beat.

It is important to note that after each smallest switch in what reality is formed into, we are back to one reality with the changed what reality is formed into integrated with the what reality is formed into that did not change. In other words, we are back to square one. What is all important is what reality is formed into and the changes that take place in what it is formed into. We always start with what reality is formed into right now. This is what can change and what we can change.

How this is actually accomplished on a universal scale defies the imagination, or at least my imagination, but that it *is* accomplished *is* here for all humans to see, experience, play around with, and enjoy!

10

A NEW WORLDVIEW

My four Reality Statements and descriptions of how reality always works everywhere combine together to give us a revolutionary new general worldview of how reality is here, changes, and continues. I call this the "One-Reformable-Reality Worldview."

Reality Generalization 1: *As any change takes place anywhere within the RF, the changed part of the RF blends together with that which did not change and becomes the RF.*
Reality Generalization 2: *There is just one continuing and changeable RF.*
Reality Generalization 3: *The same reality is used repeatedly to form every RF there ever is.*
Reality Generalization 4: *There is just one now time that is used repeatedly to give us the continuing RF we observe and experience..*

The One-Reformable-Reality Worldview is one of the same reality (defined as that which is here), with no reality added or subtracted, being used repeatedly to form every what reality is formed into there ever is. As what reality is formed into changes, what it changes to becomes what it is formed into. As these changes are taking place, continuing reality is all the reality there is. There continues to be just one what reality is formed into; it is what that one reality is formed into that can change. The same one time, which is all the time there is and has a before now and after now "thickness" of less than what we think of as being a trillionth of a second interval of time, is used repeatedly to give us the continuing reality that we observe and experience. All changes in what reality is formed into are local changes in what reality is formed into and take place in change and continuation cycles that take place in less than a trillionth of a second of marked off continuing reality.

The One-Reformable-Reality Worldview is more computer-like and less mechanical clocklike than the matter-in-motion-in-space-and-time worldview that I had previously and replaces the idea of "moving" with that of "switching."

The way reality works fools us into thinking there is a lot more reality than there really is. Having just one time that is used repeatedly and just one what reality is formed into that can be reformed repeatedly seems to me to be a "brilliant" way of accomplishing this.

The only thing that ever really happens is a minute local change or switch in what reality is formed into that takes place as a change and continuation switching cycle. We put a series of these changes together in our minds, using our memory, to form things happening.

Reality is always observed to work the general way I have described it as working, and there is nothing we can do to make it work any other way. Reality works exactly the same way regardless of whether we think the changes in what reality is formed into are "natural" (whatever that is supposed to mean) or we cause or make the changes.

All the changes that we observe taking place naturally and all of the changes that we humans can "cause" to take place are in what reality is already formed into. This is what we do when we say we are making something, or building something, or developing something. I also consider interfering with a change that is presently taking place in what reality is formed into to be "causing" a change in what reality is formed into. This is important because changes in what reality is formed into, once started, tend to repeat themselves. Regardless of what we think is happening or we are doing, what actually happens are changes within what reality is already formed into.

How does the One-Reformable-Reality Worldview differ from traditional worldviews? First of all, there are many different worldviews, perhaps as many as there human beings. Some of these are based more on stories and explanations that humans have made up that have become myths than on observations.

Those that are based on observations are often considered to be scientific. My One-Reformable-Reality Worldview differs from all previous worldviews that I am aware of in that it is about how reality itself works, which seems to be a new idea. Science has focused on how things work rather than reality itself. Science is about future predictions; the ways that reality always works are certainties. How reality always works is fundamental to how things work, and not the other way around.

Another important difference between this new worldview and previous worldviews is the way it is formulated.

I have heard it stated that the only way we can make sense of our world is in terms of space and time. Some scientific theories are formulated in terms of space and time. Matter-in-motion-in-space-and-time is certainly formulated in terms of space and time. I think I have made sense of our world without viewing or formulating it in terms of space and time, in fact, a lot more sense of it than I had when I was supposedly formulating it in terms of space and time.

I now think we have reality dimensions rather than spatial dimensions. The distance between two trees is the same regardless of whether or not there is a building between them.

Before I started making the observations that led to the One-Reformable-Reality Worldview, an important part of my worldview focused around matter in motion in space and time and Newton's three laws of motion. In this worldview, space is something apart from the physical reality of things – something that objects can move around in. In my new worldview the only space that ever exists is the one continuing space that is here along with things and is an integral part of what the one continuing reality is formed into; not something separate from it. I think Newton's laws of motion work for getting us to the moon because he was actually observing (and also measuring) what reality was formed into and changes that took place in what it was formed into. The idea of motion in space was retained even when the idea of absolute motion was dumped and replaced with that of relative motion. My new worldview dumps the idea of motion in space as being "something real" entirely. This frees our minds from a lot of superfluous thinking. Good riddance!

Another difference in my worldview from previous worldviews I am aware of is that I think the ways reality always works suggests that there is a universal super computer build into reality. I reject the idea of a "control center" outside of reality. The "outside of reality" would have to be much more complex and sophisticated that our universe. I have enough trouble trying to make sense of "our universe." Besides, if the control center created our universe, who created the control center? That would have to be even more "advanced."

New discoveries such as in genetics show that reality has an operating system and programming and other computer-like functions build into it.

I am well aware of the role that our minds play in "worldviews." With this being so, why have a worldview that focuses on what we commonly think of as being physical reality? I think, and there is much evidence to support this, that what physical reality is formed into right now (such as things and the arrangement they are in with each other) is objective and absolute. While our physical bodies including our physical brains are part of this objective and absolute, what goes on in each individual human mind is subjective.

In my new worldview, the idea that we are seeing back in time when we observe the light image from a distant star is replaced with that of seeing an image (or picture) of what some faraway part of the one continuing what reality is formed into was once formed into. Between that distance star (if it still exists right now) or a distant anything that is apart from anything else for that matter, in my new worldview there is just one continuing what reality is formed into. The idea of light traveling through space is replaced with the observable of reformations taking place in what the one continuing reality is formed into.

This removes the long held myth that, say, an arrow that is shot from a bow "literally jumps out of reality and moves freely through space." In my new worldview, the most it can "jump out of reality" is for less than a trillionth of a second of continuing reality.

The myths of the Big Bang and Expanding Space are both based on the myth of the arrow jumping out of reality and moving over an interval of time in space. And so too are a number of other myths that are now considered to be science.

My new worldview replaces the myth of a linear progressive time or for that matter any kind of passing time with that of just one reusable time. Anywhere and everywhere in the one continuing what reality is formed into the one reusable time can only be in one or the other of two states. Less than a trillionth of a second of continuing reality is not only the largest interval of time that ever exists; it is the only interval of time that ever exists. Time intervals of millions and even billions of years found in other worldviews including those considered to be "scientific" are reduced to nonsense. This removes a lot of "time" from our thinking; this is truly *Reality Simplified*!

Physics does not seem to recognize the here and now as anything at all. I now think it is everything there ever is. The value and usefulness of science seems directly related to how closely it sticks to the here and now.

I have found no evidence that there is ever any reality anywhere besides the here and now. We can remember "before" here and now and ponder here and now "after" here and now. However, the remembering and the pondering, as well as viewing old videos, is always done in the here and now.

My new worldview still fits reality-based science, including experimental science and what we think of as being "research studies." Thus, it is useful not only for daily living, but also for doing science.

My new worldview is more realistic than any that have come before it.

You now know what the One–Reformable-Reality Worldview is and how it differs from other worldviews. The remainder of this book focuses on how this revolutionary new worldview can be used for living better and more meaningful lives as well as for making our world a better place to live.

11

LIVING IN THE ONE CONTINUING REALITY

When I started making the observations that led to the One-Reformable-Reality Worldview, my goal was to figure out the general way that reality is here, changes, and continues. I thought this might result in better understanding of our physical world, but this would be a mind thing. I had no idea that it would be useful in the actual living of our lives. I was surprised when I actually had my new worldview housed in my mind that this somehow drastically changed my life for the better, not only for manipulating the world to my advantage, but also for deciding what I should and should not change and do.

In discussing the utility of my new One-Reformable-Reality Worldview, I will begin with the Planet Earth, which because it provides both a place and resources for living our lives is at the top of the list for humans of the things we find reality formed into right now. It is obvious, however, that the earth is a continuing thing, subject to change, rather than a "permanent object."

The Planet Earth is a continuing thing with a high degree of stability within what the one continuing reality is formed into, but is subject to changing. Hopefully, it won't change so much that we would no longer consider it to be the Planet Earth, or so much that it would not be a continuing place for us to live our lives. In any case, it is not a permanent object. Continuing reality is all the reality the Planet Earth ever has.

Once we realize that our lives are taking place in and as an integral part of the one continuing what reality is formed into, we can use this knowledge for living better and more meaningful lives. At the heart of the way our lives take place is the reforming of what reality is already formed into, which becomes what reality is formed into over and over again.

The only thing that can change or be changed is what the one continuing reality is formed into right now. And like magic, any change that takes place or we "cause" to take place in what the one continuing reality is formed into blends together with that which did not change and becomes what the one continuing reality is formed into. I for one never get tired of playing around with this amazing way that reality always works..

This is the given that allows me to brush my teeth, tie my shoes, cook a meal, and anything else I can possibly do, including walk around.

I once thought I built a boat from "scratch," but what I actually did was reform parts of what reality was already formed into to a boat.

Whenever we say we are building or developing or inventing or making something, what we actually do is reform some of what reality is already formed into to something else. We can never build anything from "scratch." We always start with some of what reality is already formed into. Realizing this has helped me to live a better and more creative life. Remember, we always begin with what reality is formed into right now.

It is our ability to reform what our own bodies, including our physical brains, are formed into that allows us to reform what the one continuing reality around us is formed into and also allows us to walk about and change where we are at in the one continuing what reality is formed into. We can thus move about within what reality is formed into to interesting and fascinating parts of what it is formed into. We can also accomplish this reforming within what reality is already formed into by riding a bicycle or driving an automobile. Once we realize this, we can do it more creatively.

Regardless of what we think is happening or we are doing, it takes place as the reforming of what reality is already formed into. It is the one continuing what reality is formed into that is absolute and objective; what we think is happening or we are doing is subjective. It's based on differences in what we remember the one-continuing what reality is formed into compared to what it is formed into right now, and our memories are subjective. Photos and video recordings often give a more "objective" picture of what the one-continuing reality was once formed into that can be used to try to figure out "what really happened." What we say happened often does not agree with what video cameras show.

I will now focus on how we use our here and now. There is a recent surge of published self-help writings about the importance of living in the present moment. Much of this comes from Eastern philosophy. The essence of all this is that we should live in the present moment, rather than thinking about the past (which has already happened and cannot be changed) and/or imagining the future, such as what might happen tomorrow or next year and then worrying about that.

The problem with this is that according to my One-Reformable-Reality Worldview, our lives take place entirely in the here and now. I have concluded that this is true because the here and now is all that ever exists; that there is never any reality besides this. This brings to light that I live my life in a continuing right now rather than a passing time, whatever that is supposed to be. I don't have any choice.

This being said, it then becomes apparent that I can only do something, including mind things like thinking about the past or imagining the future in the continuing here and now. Even a video of past here and now can only be viewed starting at a right now in the continuing here and now.

So it boils down to how we use our here and now. I love new pleasant experiences and adventures and these involve actively living in the present moment. My RVing allows me to do this in exciting and unique places. We can never actually live in the past or future, but we can use some of our continuing here and now to recall past here and now and imagine and plan for what future here and now might be like. This might lead to worry and anxiety about what we have done in the past or what we imagine might happen in the future. But we can also use this thinking to learn from our past mistakes and to plan for a better future.

The importance and possibilities of using some of our continuing right now for creative mind adventures and problem solving is often overlooked. The self-help writings suggest that it is beneficial to be fully aware of the present moment, such as by using meditation to focus on our breathing. When you take a walk in a park, be aware of your body movement and footsteps and take in the sights (such as the leaves on the trees) and sounds (such as the birds make) and smells (such as of the cottonwood blossoms) and feel the heat of the sun. I sometimes try to do this, and it can be pleasant being engulfed in the present moment. But it can also be pleasing to be lost in my thoughts as I walk along, churning ideas around in my mind, thinking about how I can bring the chapter I am working on in a book I am writing alive. I think of these as mind journeys. It was the longest and most focused mind journey that I have made so far that led to this book.

What one person thinks is wasting here and now, such as by watching television or other things that take place on screens or deadening our awareness of the continuing here and now with alcohol and drugs; others might think is using the here and now wisely.

To each his or her own, but I prefer to use my continuing here and now actively rather than passively, regardless of whether it takes place in the physical world or is a journey in my mind. Both can be equally rewarding. As for worry and anxiety over thinking about the past or planning for the future, I suggest you replace the idea of passing time with that of continuing reality. Most of the worry and anxiety is over "passing time" that does not exist. It's right now right now, and it's right now right now. Always has been and always will be. This is one thing you don't have to worry about.

I now know that my life takes place entirely in continuing reality. I can only view a video of my past in continuing reality as reality is continuing.

Instead of thinking "time," I now think "continuing reality." Instead of thinking clocks tell me what time it is, I now think clocks mark off points I am at in the continuing reality of today.

My past took place in continuing reality. Right now, my life is taking place in continuing reality. My future will take place in continuing reality.

I think about my past and plan for my future, but I do this thinking and planning in continuing reality. All that remains of my past life including all memories I and other people have of it and all recordings and other memorabilia are being carried along in continuing reality right now.

The past should be thought of as past continuing reality. The future should be thought of as future continuing reality.

Everything that happens does so in the one continuing what reality is formed into. This includes everything that anybody, including a scientist, can observe happening or can make happen.

To reform or not to reform – to leave it as it is or change it – is a choice we often need to make.

12

MAKING OUR WORLD
A BETTER PLACE

The One-Reformable-Reality Worldview gives us a new view of history. Our future is in the reforming of what reality is formed into, and so too was our past. What the one continuing what reality is formed into right now is absolute and objective. This is the result of all the natural reforming that has taken place in the past and the reforming humankind has added to this. This is the objective and absolute history rather than the subjective one made up in human minds.

We can look back by means of memories, recordings, and measurements at what the Planet Earth was once formed into at a certain amount of continuing reality ago and compare that with what, say, the same place on the surface of the earth is formed into now.

This gives a different view of climate change than does matter in motion in space and time myth. What reality is formed into right now is objective and changeable and what we should be concerned about.

Climate change is only one part of the story of what is our environment and the Planet Earth are being reformed into. Even more potentially catastrophic is that we are overpopulating our planet and trashing it. What this amounts to is that we humans are overall reforming the earth into something less desirable. The consumer society is literally wasting away our natural resources by turning a lot of them into junk and garbage. Does anyone think this is progress? To a large measure this is brought about by a misunderstanding of how reality always works. In simple terms, all that we can ever cause to happen to our environment and the Planet Earth is the reforming of what it is already formed into, which becomes what it is reformed into, perpetually, over and over again. A useful and simplified way of thinking about and observing what reality is already formed into is as the specific things or stuff it is formed into and the arrangement the specific things or stuff is in with each other. Natural changes and cycles work the same way. They cause what reality is already formed into to change and become what it is formed into. Ultimately, we do not know why it works this way. But we can observe that it does work this way.

Humankind seems to have one idea of what reality should be formed into and nature another, and the two, rather than working together, seem to be in a battle with each other. We humans seek dominance over nature.

Global warming seems to be a natural cycle of nature causing changes in what reality is formed into that is now taking place. However, our bad habits are making it much worse than it would otherwise be. In other words, we are greatly speeding it along. What we should be doing is to try to keep the natural cycle of climate change as close to natural as humanly possible. This is the goal of the environmentalists. Simultaneously, we need to continue to live our lives.

How can we balance all of this out? It is an understatement to say that this is a complex problem. However, it is obvious that we can do this better than we are doing it now.

Humans now have tremendous power to reform what reality is already formed into. We literally have the power to move mountains with heavy-duty earth moving equipment and explosives. But just because we have the power to do something or "develop" something does not necessarily mean we should do it.

Right now as I write this, is the reforming of what reality is already formed into going in the right direction? At least in some cases we are slowing down reforming that is going in the wrong direction. However, we can and hopefully will do better and make our world a better place to live.

My concern right now as I write this is that there is a general trend toward throwing the future of the one-continuing RF by the wayside and raping what is left of our natural resources for a last big boom? That would certainly be reforming what reality is already reformed into in the wrong directions. Let us hope that won't happen. The future is in the reforming. It's not too late to make our world a better place to live.

ABOUT THE AUTHOR

Jack Wiley earned his Ph.D. at the University of Illinois in 1968 and then did research at three different universities for a total of four years before going off on his own so that he could do his own thinking. Dr. Wiley's revolutionary new worldview is the result of that thinking.

For more information about the author and his books, go to: **http://www.amazon.com/author/jackwileypublications.**